Why Does J Reject Jesus

Unintended Messianic expectations and unplanned temple-worship

Dr. Ali Ansarifar

Kingdom Publishers

www.kingdompublishers.co.uk

Why Does Judaism Reject Jesus Christ?

A catalogue record for this book is available from the British Library.

All Scripture Quotations have been taken from the Interlinear Version of the Bible

ISBN: 978-1-911697-73-2

1st Edition 2023 by Kingdom Publishers
Kingdom Publishers

London, UK.

You can purchase copies of this book from any leading bookstore or email
contact@kingdompublishers.co.uk

Preface

The history of God's engagement with His people, the Israelites, is recorded in the Old Testament of the Bible. It is said that God delivered the Israelites from slavery with the intention of transforming them into a model nation by commanding them to obey strict monotheism and His moral law. However, this did not happen, and when the history of God's engagement with the Israelites is examined closely, a disturbing pattern emerges. After the Israelites were given the Ten Commandments, God embarked them on conquering the land to settle the twelve tribes of Israel. As the years passed, kings and kingdoms appeared in the national life of Israel, and temple-worship and blood sacrifices took hold and became the traditional way of life and worship for the Israelites for centuries to come. This happened despite God's disapproval. Israel's history, then, fell into a pattern of constant conflict through invasion by foreign armies, expulsion from the conquered lands, and the diaspora. The Temple in Jerusalem was destroyed twice, and the people were driven into exile and returned in the twentieth century to create the state of Israel.

This book proposes that the history of ancient Israel reflects a fundamental problem with the human soul, a malignancy that rejects strict monotheism and resorts to paganism in the form of idolatry and improper conduct. The actions that the Israelites took alienated them from God and set them on a course to catastrophe. A close examination of some of the verses in the Old Testament reveals the exact nature of this malice, which is discussed in depth by the author. The aim is to reiterate that there is a Solution; redemption, to cure this malignancy and restore humankind's relationship with God. The alternative to redemption is eternal separation, death, and even damnation. The question is whether humanity will seize this moment of grace and accept God's call to repentance or continue on a path to destruction.

Dedication

Those who seek and find the biblical truth and share it with others for the redemption of humankind and the glory of God through Jesus Christ the Saviour

Disclaimer

It has never been the intention of the author of this book to infringe on the sensitivity, personal faith, or belief of his readers. The material and information presented in this book are never meant to cause offence to any individual person or group of people of any race, creed, nationality, or background. This book is written for educational purposes and academic interest only.

Jesus said, "Whatever you hear with your ear, proclaim upon your rooftops into the other ear. Indeed, no one lights a lamp and puts it under a vessel, nor puts it in a hidden place. Rather it is put on a lampstand so that each who enters, and leaves might see its light." (Gospel of Thomas 33) [1]

When we strip away the worthless knowledge and wisdom that the world has given us and trash them without fear, we will then become the children of the living God and receive the truth.

Neither the Holy God nor ancient Israel ever planned for a Messianic figure or temple-worship. While God saw the salvation of Israel in the purification of its inner heart, Israel believed its salvation was in the destruction of its mortal enemies. God's anointed Saviour, Jesus Christ, cleanses the heart, but what benefit does it have for Israel, which faces destruction by its gentile enemies?

The story of ancient Israel is like this.
A farmer planted a seed in the ground. The seed
grew into a tree, but the tree produced no fruit.
When the seed died, the die was cast, and the
farmer had no say.

As it was with Moses, Jesus was the holy ground on which God stood and revealed Himself to humankind. But when God relinquished the holy ground, Jesus cried, "Father, why have you forsaken me?"

Why did God forsake Jesus on the cross?

The human soul is like soil. It never produces exactly what is expected when it is planted, but one must make do with the harvest it yields, as little as it may be.
Blessed are those whose inner hearts seek God the Father. For their sakes, the truth was revealed in Jesus Christ.

Every man is like farmland. When the land is inspected, the farmer says, "How disappointing that the harvest is so poor. If only I had a say when the land was formed."

Contents

Chapter 1
A model nation and a priesthood –
redemption of the gentiles

The Bible provides a comprehensive record of God's revelations to a people who were freed from slavery in Egypt and brought into a new land to start a fresh chapter in their history. God planned to make them into a model nation, referred to as Israel, to redeem the gentiles. But their history took a different course from the one God intended and changed the entire scheme for human salvation. The events that ensued their departure from slavery in Egypt had a profound impact on their lives and the destiny of humankind.

The seed of Israel - Long ago and far away, a family worshipped idols. The son, Abram, said to his father, "What help and profit have we from those idols which you worship, and in the presence of which you bow yourself? There is no spirit in them. They are dumb forms, and they mislead the heart. Do not worship them, worship the God of heaven, who causes the rain and the dew to fall on the earth and does everything on the earth, and has created everything by His word, and all life is from His presence." [2] Abram married Sarai, and she became his wife. One night, Abram observed the stars from the evening to the morning, and a word came into his heart, and he said, "All the signs of the stars, and the signs of the moon, and of the sun, are all in the hand of the Lord. Why do I search them out? If He desires, He causes it to rain, morning and evening, and if He desires, He withholds it, and all things are in his hand." He prayed in the night and said, "My God, God Most High, You alone are my God, and You and your dominion have I chosen. And You have created all things and all things that are the work of your hands. Deliver me from the hands of evil spirits who have dominion over the thoughts of men's hearts, and let them not lead me astray from You, my God. And establish me and my offspring forever so that we do not go astray from now and forever." Then the word of the Lord was sent to him through an angel, saying, "Get out of your country, and from your kindred, and from the house of your father, and go to a land which I will show you, and I shall make you a great and numerous nation. And I will bless you and I will make your name great, and you will be blessed in the earth, and in You shall all families of the earth be blessed, and I will bless them that bless you, and curse them that curse you. I will be a God to you and your son, and to your son's son, and to all your offspring. Fear not, from now on and to all generations of the earth, I am your God." God changed Abram's name to Abraham and his wife's name to Sarah. [2]

Abraham and Sarah had a son and named him Isaac. The descendants of Isaac's son, Jacob, referred to as Israelites, were slaves in Egypt for many years. God judged the Egyptians and inflicted punishment on them. He then brought the Israelites out of Egypt through His servant Moses and his brother Aaron and took them to a rich and fertile land, as He solemnly promised to their ancestors. The Israelites came to the desert of Sinai and set up camp at the foot of Mount Sinai. The Lord called to Moses from the mountain and told him to say to the Israelites: "You have seen what I did to Egypt, and I bore you on the wings of eagles and brought you to Me. And now if you will surely listen to My voice, and will keep My covenant, you shall become a special treasure to Me above all the nations, for all the earth is Mine. And you shall become a kingdom of priests for Me, a holy nation. These are the words which you shall speak to the sons of Israel." And Moses came and called the elders of the people. And he put all these words before them which Jehovah commanded him. And all the people answered together and said, "All which Jehovah has spoken, we will do." And Moses brought back the words of the people to Jehovah. (Exodus 19:4-8) [3] And Jehovah said to Moses, "Go to the people and sanctify them today and tomorrow. And let me wash their clothes. And be ready for the third day. For on the third day, Jehovah will go down before the eyes of all the people on the mountain of Sinai. And you shall set limits to the people all around you, saying, "Be careful with yourselves, not going up into the mountain and touching its border – everyone touching the mountain shall surely be killed.

Not a hand shall touch him, but surely, he shall be stoned, or surely he shall be shot through." (Exodus19:10-13) [3] The mountain became a holy place for God to stand on and reveal Himself to His people. God spoke. "I am Jehovah your God, who has brought you out from the land of Egypt, from the house of bondage." Jehovah then gave the ten commandments, or the moral law, to the Israelites through Moses and Aaron. The first three commandments forbade the Israelites from having any other gods besides Jehovah, from making idols, from serving or worshipping idols, and from using the name of Jehovah in vain. They were given to create a monotheistic people who would serve God as His priests, a model nation. The fourth and fifth commandments demanded observance of the Sabbath (a day of rest dedicated to Jehovah) and respect for one's father and mother. The latter were supposed to provide a day of rest from work and a peaceful and loving family home. The remaining commandments demanded the avoidance of murder, adultery, theft, false accusation, and coveting another man's possessions and were meant to provide a guideline for moral conduct among the Israelites.

God's promises and instructions to Israel - The Lord said, "Behold, I am about to send an angel before you, to guard you in the way, and to bring you to the place which I have prepared. Be observant before Him and listen to His voice. Do not be rebellious against Him, for He will not forgive your transgressions; for My name is in Him. For if you fully listen to His voice, and do all which I speak, I will be an enemy to your enemies. You shall not bow down to their gods, and you shall not serve them. And you shall not do according to their works. But you shall surely tear them down, and you shall surely smash their pillars. And you shall serve Jehovah your God, and He will bless your bread and your water. And I will remove sickness from your midst. There shall not be one miscarrying, nor one barren in your land. I will send my terror before you, and I will confound all the people among whom you come. And I will among whom you come; give the neck of your enemies to you. I will drive your enemies out before you little by little until you are fruitful and possess the land. For I will give the people of the land into your hand. And you shall drive them out before you. You shall not cut a covenant for them and for their gods. They shall not dwell in your land, lest they cause you to sin towards Me. Surely when you serve their gods, it will be a snare to you." (Exodus 23:20-33) [3]

Then Moses told all the words of Jehovah to the people and all the judgements. And all the people answered with one voice and said, "We will do all the words which Jehovah has spoken." And Moses wrote all the words of Jehovah. And he rose early in the morning and built an altar below the mountain and twelve memorial pillars for the twelve tribes of Israel. And Moses took half of the blood of bullocks, and he put it in basins. And he sprinkled half of the blood on the altar. And he took the book of the covenant and read it in the ears of the people. And they said, "We will do all that Jehovah has spoken, and we will hear." And Moses took the blood and sprinkled it on the people, and said, "Behold, the blood of the covenant which Jehovah has cut with you concerning these words." (Exodus 24:1-9) [3]

The covenant that God made with ancient Israel demanded strict monotheism and full obedience to His moral law. Strict monotheism ensured reliance on God and the moral law; the good conduct of a model nation or priesthood worthy to serve the Holy God.

God's warnings to Israel – God issued specific warnings to Israel. They were instructed not to create idols or erect graven images of any kind, to observe His Sabbaths and revere His Holy Place, and to keep and do His commandments. In return, God promised to live and walk among them, to keep their land fertile and plentiful; to secure it from internal and external enemies; and to make it peaceful so they could multiply and become a numerous nation. God would live among them forever. But if Israel abandoned its God and resorted to idolatry and immorality, then God would appoint terror over them, destroy their health and wealth, and abandon them to their enemies and those who hate them. If Israel continued in its sinful ways, God would chastise it and break its pride and strength; remove His blessings from the land; bring plagues and hunger; and a sword on it, carrying out the covenant's vengeance; make its cities desolate; and scatter its population among hostile nations, drawing the sword after them. And those who remain will putrefy in the iniquities of their fathers and of their own. (Leviticus 26:1-40) [3] One prophetic warning in Exodus 26:14 is of special interest here. "And if you will not listen to Me, and do not do all these commandments; and if you reject My statutes, and if your soul hates My judgments, so as not to do all my commands, to the breaking of My covenant, I will do this to you also, and I shall appoint terror over you, the burning fever, destroying the eyes and consuming the soul." (Leviticus 26:14-16) [3] This is a clear indication that God, by nature, is very authoritarian and strict. There was no get-out clause in the covenant after the Israelites were sworn in.

Summary – The plan for the redemption of the gentiles was centred on a model nation or a priesthood that would serve God. The covenant made with ancient Israel had two major components: strict monotheism and obedience to moral law. In this covenant, Israel's devotion to God through strict monotheism and adherence to His moral law through the ten commandments were paramount. God promised to develop and safeguard Israel into a unique model nation as an example to the gentiles, and by doing so, transform the pagan ways of the gentiles into godliness and perfect morality. The covenant demanded loyalty to God and obedience to His moral law by reminding the Israelites of the fate of their ancestors who were brought out of slavery in Egypt. However, over time, this proved insufficient to keep the Israelites on a godly path.

Conclusions

- God is authoritarian and demands full loyalty and compliance with His moral law.

- There was no Messianic figure or a promise of a future Messiah in the original covenant.

- The acquisition and transfer of land from the gentiles to the Israelites were conditional. So long as the Israelites remained faithful to God and obeyed His moral law, they could have enjoyed ownership of the land. If the terms set out in the covenant were violated, the covenant would have been nullified and the ownership of the land would have come to an end.

Chapter 2
An embryonic Messiah – heroes and military leaders of Israel

There was a violent and lawless period in Israel's history when God orchestrated military campaigns and appointed military leaders to invade, conquer, kill, and remove the gentiles and assign the captured territories to the tribes of Israel. However, as military campaigns proceeded, frequent idolatry and immorality caused Israel's relationship with God to deteriorate. One unintended consequence was the rise of national heroes and military leaders who supplanted God as His people's sole saviour. A national saviour, a Messiah, who would defeat Israel's enemies and secure its national borders was engraved in the psyche of Israel.

The turbulent history of Israel – In Israel's laborious and troublesome history, some events and places changed the destiny of God's people. After Moses died, the conquest of the gentile land began in earnest through military campaigns and the capture of territories for the purpose of assigning them to the tribes of Israel. As time passed, a lawless period in Israel's history began and continued until the coming of Jesus Christ. This period of Israel's history was extremely violent and had unintended consequences for generations. God orchestrated a military campaign to capture land from the gentiles and tear down their altars and destroy their pagan gods before resettling the tribes of Israel. God was fulfilling the promise He made to Jacob to give land to his descendants to multiply and become a numerous nation. However, this process was marred by idolatry and immorality and led to profound changes in God's relationship with His people. Some verses in the book of Judges provide a graphic description of a terminal malignancy in the soul of ancient Israel that led to idolatry and improper conduct. On one occasion, the angel of Jehovah said, "I caused you to come up out of Egypt and brought you into the land which I had sworn to your fathers, and said, I shall not break My covenant with you forever. And you, you shall cut no covenant with the inhabitants of this land . You shall break down their altars. Yet you have not listened to My voice. What is this you have done?" And I also said, "I shall not drive them out before you, and they shall become adversaries to you; and their gods shall become a snare to you." (Judges 2:1-3) [3] God's people served their appointed leaders and the elders, who saw all the great work of Jehovah which he had done for Israel. But when the great leaders died, and another generation arose, after them who had not known Jehovah, nor yet the works which He had done for Israel, the sons of Israel did evil in the sight of Jehovah and committed idolatry by going after other gods, of the gods of peoples around them, and bow themselves to them, and angered Jehovah. And the anger of Jehovah burned against Israel, and He gave them into the hand of plunderers; and they plundered them. And He sold them into the hands of their enemies all around them, and they were not able to stand before their enemies any longer. Whenever they went, the hand of Jehovah was against them for evil, as Jehovah had spoken and as Jehovah has sworn to them. And it distressed them very much. And Jehovah raised up judges, and they saved them from the hands of their plunderers. But they also did not listen to their judges, but went whoring after other gods, and bowed themselves to them. They quickly turned aside out of the way in which their fathers walked to obey the commands of Jehovah. They did not do so. And when Jehovah raised up Judges to them, then Jehovah was with the Judge, and rescued them out of the land of their enemies all the days of the Judge. For Jehovah took pity because of their groaning before the oppressors, and those that crushed them. And at the death of the Judge, it happened that they would turn and act more corruptly than their fathers, to go after other gods, to serve them, and to bow themselves to them. And they did not fall away from their own doings, and from their stubborn way. And the anger of Jehovah glowed against Israel. And He said, "Because the nation has transgressed My covenant which I commanded their fathers, and has not listened to My voice, I also from now on will not drive out any from before them, of the nations before Joshua left when he died; so that by them I may test Israel, whether they are keeping the way of Jehovah, to go in it, as their fathers kept it, or not. And Jehovah left those nations without driving them out quickly. And He did not deliver them into the hands of Joshua." (Judges 2:6-13) [3] As God's displeasure with the conduct of Israel grew, He relied more on military leaders to rescue Israel during adversity and national emergency. God became an advisor on military strategy and appointed leaders to wage wars for and defend Israel against its enemies. In a time of distress, the Israelites cried out to God for help, promising faithfulness and compliance with His moral law. However, after the Lord delivered them from their predicament, they reverted to their old sinful ways. This became a reoccurring cycle of events in ancient Israel's history, and God's relationship with His people never fully recovered. This period of Israel's history was critical in shaping future events and expectations. The bloody wars that Israel waged to capture land violated the moral law that states, "You shall not murder"; "You shall not covet your neighbour's house"; "You shall not covet your neighbour's wife, or his male slave, or his slave girl, or his ox, or his ass, or anything which belongs to your neighbour." (Exodus 20:13&17) [3] This violation was of no concern because the Lord God sanctioned it. Still, other breaches of the moral law were strictly prohibited and treated as sin. Furthermore, as more land was conquered, mixed marriages between the Israelites and gentile women became more common, displeasing God.

Summary – God's covenant with the Israelites demanded strict monotheism and obedience to His moral law. However, when the campaign to secure land for the tribes of Israel began, the Israelites encountered the pagan cultures of the gentiles. As a result, idolatry and improper conduct became more widespread among God's people, nullifying the covenant with collective Israel. As the campaign continued, heroes and military leaders displaced God over Israel's affairs in their wars of conquest and were able to secure victories for Israel against its enemies. The unintended consequence was the emergence of a national saviour, a Messianic figure, who would save Israel from its enemies during adversity. A Messianic figure was engraved in the psyche of Israel. The subsequent events and turmoil that ensued in the history of God's people in the holy land after establishing the kingdoms of Israel and Judaea defined this Messianic figure as a military hero. He saved Israel during national emergencies. Neither the Holy God in His original covenant with ancient Israel nor the Israelites ever planned for a Messianic figure. While God saw the salvation of Israel in purifying its inner heart, Israel believed its salvation was in the destruction of its mortal enemies through heroic military leaders. God's anointed Saviour, Jesus Christ, cleanses the soul, but what benefit does it have for Israel, which faces destruction by its gentile enemies? This period in Israel's history fundamentally departed from God's original plan to secure land and create a model nation or priesthood to redeem the gentiles. As the Israelites continued violating the covenant, a cycle of retribution and punishment took hold, and God's relationship with His people never fully recovered. Soon after the Israelites were established in the holy land, Israel moved away from God and tilted towards the gentile world, where kings and kingdoms ruled the people's affairs and destiny, not the Holy God. At this crucial time, the die was cast, and God's salvation plan for the gentiles through Israel failed. In effect, the original covenant was dead. Heroic military figures such as Joshua, Gideon, Samuel, and Samson were imprinted in the imaginations of God's people for centuries to come.

Conclusions

- The covenant God made with ancient Israel came into jeopardy during the military conquest of land to resettle His people.

- In this period, an unintended Messianic expectation emerged in the psyche of Israel, the embryonic form of which was heroic military leaders who would save Israel during national emergencies. This embryonic form of Messiah would develop further throughout Israel's turbulent history to replace God's Messianic figure, Jesus Christ, who was humble and appeasing. While God saw His people's salvation in purifying their hearts, they saw it in defeating their mortal enemies and securing their national borders. The scene was set for the rejection of God's Messiah.

• Unintended Messianic expectations were the legacy of this period in Israel's history and they provided the first building block of Jewish history.

The next chapter will discuss Israel's historical development into two kingdoms and the immense influence the lives and achievements of the kings had on the affairs of and future of their subjects.

Chapter 3
Judaism's coming of age – the historical structure of Judaism.

As the Israelites settled down in the holy land, Israel advanced from the period of heroic military leaders to the monarchy, when kings and kingdoms ruled the land. God reluctantly appointed kings for Israel, and the people and the kings remained under the sovereignty and judgement of God. However, the kings often struggled with internal and external enemies, and despite their deep faith in and devotion to God, their desires and ambitions led them to disobey God's moral law and sin. The lives of these kings were so impressive that the people of Israel longed for another king. As the line of kings was firmly established, God was side-lined. In this period, temple-centred worship became a part of Israel's religious devotion and national life.

Who should be your king? – In the original covenant and through the first commandment, God explicitly asked for the full devotion of His people and regarded Himself as the real King of Israel. When the people asked for a king, Samuel, God's servant, gave this warning. This is how your king will treat you. He will make soldiers of your sons;

some of them will serve in his war chariots, others in his cavalry, and others will run before his chariots. He will make some of them officers in charge of a thousand men, and others in charge of fifty men. Your sons will have to plough his fields, harvest his crops, and make his weapons and the equipment for his chariots. Your daughters will have to make perfumes for him and work as his cooks and his bakers. He will take your best fields, vineyards, and olive-groves, and give them to his officials. He will take a tenth of your corn and of your grapes for his court officers and other officials. He will take your servants and your best cattle and donkeys and make them work for him. He will take a tenth of your flocks. And you yourselves will become his slaves. When that time comes, you will complain bitterly because of your king, whom you yourselves chose, but the Lord will not listen to your complaints. The people paid no attention and said, "We want a king, so that we will be like other nations, with our own king to rule us and to lead us out to war and to fight our battles." The Lord gave them a king. (1 Samuel 8:10-22) [3] In response to the people's request, God chose a king for them. In effect, the Israelites rejected the kingship of God and began to look more like the gentile nations around them. The idea of a model nation entirely devoted to God through strict monotheism and obedience to His moral law ended after Israel made its choice despite God's displeasure. God was no longer a source of their existence, devotion, strength, guidance, protection, and inspiration for most people, but the kings were. At this critical time in Israel's history, God was no longer the king of His people, and they chose to be like the gentiles rather than become a model nation as He intended. Any hope of revitalising the covenant God made with their ancestors was dashed. On the other hand, God continued to be faithful to those who called on His name and upheld His moral law. Verses 10-22 of Samuel 8 provide an accurate description of the power structure and methods that the empires and the ruling classes have used throughout history to govern and exploit human societies. The covenant meant creating a model nation to break this never-ending cycle of exploitation, violence, injustice, instability, theft, class struggle, and bloodshed; freeing humans from slavery; restoring their self-worth and dignity; and bringing them to the light of the Father God. Sadly, it never happened, and the relationship of

God's people with their gentile neighbours sometimes remained troublesome and violent.

Samuel gathered the people and said, The Lord, the God of Israel, says, "I brought you out of Egypt and rescued you from the Egyptians and all the other people who were oppressing you. I am your God, the one who rescues you from all your troubles and difficulties, but today you have rejected me and have asked me to give you a king. Very well then, gather yourselves before the Lord by tribes and by clans." God then chose a king for Israel. " Now here is the king you chose; you asked for him, and now the Lord has given him to you. All will go well with you if you honour the Lord your God, serve him, listen to him, and obey his commands, and if you and your king follow him. But if you do not listen to the Lord but disobey his commands, he will be against you and your king. So then, stand where you are, and you will see the great thing which the Lord is going to do. It's the dry season, isn't it? But I will pray, and the Lord will send thunder and rain. When this happens, you will realize that you committed a great sin against the Lord when you asked him for a king."(1 Samuel 12:13-17) [3]

It is puzzling why, despite the Israelites' frequent disobedience to His moral law, God expected them to listen to their kings and obey their commands. Some kings were profoundly corrupt and sinful and were not good examples to follow. Moreover, if the Israelites did not follow God's commands, then how were they expected to follow their king's commands and remain law-abiding?

There were conflicts among God's people and frequent military defeats when the kings ruled. Betrayal, murder, rebellion, and revenge were common. In one battle, the Israelites were defeated and brought in the Lord's Covenant Box to help them win the battle. Unfortunately, the Covenant Box was captured after Israel was defeated and eventually returned. (1 Samuel 4:1-10) [3] By this time, the Israelites were stealing religious relics to help them win their wars, which had little or no benefit and was evidence of further deterioration in their relationship with God.

The great king and the temple - Israel had many kings, but Solomon (970–931 BCE) was probably the greatest of them. Solomon asked God for wisdom to rule over his people and God granted him more wisdom and understanding than anyone has ever had before or will ever have again. God also bestowed upon Solomon greater wealth, honour, and longevity than any other king, provided he obeyed and followed His moral law and commands. (1 Kings 3:10-14) [3] Solomon ruled over the kingdoms of Judah and Israel and some neighbouring countries. The people throughout the land were prosperous and lived in safety and peace. Solomon's wisdom proceeded him, and he composed proverbs and more than a thousand songs. His knowledge of trees and plants, animals, birds, reptiles, and fish were so impressive that kings all over the world heard of his wisdom and sent people to listen to him. (1 Kings 4:20-33) [3] Then Solomon remembered the promise God made to his father, King David. "Your son whom I will appoint in your place on your throne, he shall build the house for My name."

Solomon decided to build that temple for the worship of the Lord. (1 Kings 5:5) [3] In the original covenant, there was no mention of a temple for worshipping the Lord, and the word "worship" never appears in the ten commandments. Moreover, David's last instructions to Solomon before he died were, "To keep the charge of Jehovah your God, to walk in His ways, to keep His statutes, His commands, His judgements, and His testimonies, as it is written in the laws of Moses." (1 Kings 2:3) [3] King David was living in his palace, built of cedar. The Covenant Box was kept in a tent, so David decided that the Lord should also live in a palace like his. The Lord said, "Go, and you shall say to My servant David, So says Jehovah, You shall not build a house for Me to dwell in; for I have not dwelled in a house from the day that I brought Israel up until this day, but I have gone from tent to tent, and from one tabernacle to another." (1 Chronicles 17:4-5) [3] There was no mention of building a temple for worship in the original covenant with ancient Israel, and furthermore, God preferred to live in a tent. But at the insistence of King David, God reluctantly agreed to have a temple built and asked that Solomon should remain loyal and law-abiding. The Lord said to Solomon, "As to the house that you are building, if you shall walk in my statutes and shall do My judgements; you shall do all my commandments, to walk in them, then I shall establish your word with you, which I spoke to your father David. And I shall dwell in the midst of the sons of Israel and shall not forsake My people Israel." (1 Kings 6:12-13) [3] Solomon also built a palace for himself.

The Temple was constructed, the ark of Jehovah, the tabernacle of the congregation, and all the holy vessels that were in the tabernacle, even those of the priests and the Levites, were brought in. And King Solomon and all the company of Israel who had assembled for him were with him before the ark, where the two tablets of stone which Moses put there, sacrificing sheep and oxen. The cloud of the glory of Jehovah filled the house of Jehovah. (1 Kings 8:4-5,11) [3] Solomon addressed his people and said, "And it was in the heart of my father David to build a house for the name of Jehovah, God of Israel; and Jehovah said to my father David. Because it has been in your heart to build a house for My name, you have done well that it has been in your heart; only, you shall not build the house, but your son who shall come out from your lions, he shall build the house for My name." (1 Kings 8:17-19) [3] The Lord then appeared to Solomon and said, "I have heard your prayers and your supplication with which you have made supplication before me; I have hallowed this house that you have built; put My name there forever; and My eyes and My heart shall be there forever. And you, if you walk before me as your father David walked, in singleness of heart and in uprightness, to do according to all that I have commanded you – you shall keep My statutes and My judgement – then I shall establish the throne of your kingdom over Israel forever. If you at all turn back, you and your sons, from following Me, and do not keep My commands, My statutes which I have set before you, and you shall go and serve other gods and bow yourselves to them, then I shall cut off Israel from the face of the land that I have given to them, and the house that I have hallowed for My name I shall send away from My presence, and Israel shall be a proverb and a byword among all people." (1 Kings 9:1-6) [3] Despite God's displeasure and repeated warnings, Solomon's incredible appetite for non-Hebrew women was his downfall. As

Solomon intermarried with foreign women, they made him turn away from God, and in his old age, he built places of worship where he and his wives burned incense and offered sacrifices to their gods. God became enraged and took away his kingdom. After Solomon died, Israel's golden age ended abruptly, and the land was ruled by corrupt kings and consumed by conflict and bloodshed. The kingdom of Israel was destroyed after falling into pagan worship and immorality, but the kingdom of Judaea survived.

Summary - When the kings ruled, and kingdoms shaped the national life of Israel, this was a turning point. The Israelites asked for their own king to rule over them, lead them to war, and fight their battles like other nations. The Lord reluctantly gave them a king, and God's kingship over Israel ended. But their desire to become like other nations would never be realized, and their history would remain in constant conflict with the gentiles for centuries to come. Temple-centred worship and blood sacrifices were not included in the original covenant. Instead, they were added to Israel's national life due to the king's inspiration to please God. The unplanned temple worship and blood sacrifices became the second building block, and together with the unintended messianic expectations (Chapter 2), they formed the structure of Judaism in the ensuing centuries. The rigid structure of Jewish history was firmly in place, and the destiny of Israel was no longer in God's hands. As the era of the great kings ended, Israel's history and fate evolved around the Temple in Jerusalem. God relinquished His sovereignty over Israel, but He remained faithful to those who trusted Him and followed His moral law.

Conclusions

- No temple-centred worship or blood sacrifices were mentioned in God's original covenant with ancient Israel, and God expressed His disapproval of the Temple.

- However, these rituals, together with the Messianic expectations, were added during the rule of Israel's military heroes and kings and formed the pillars of Jewish history in the ensuing centuries, long after Israel's kings were gone.

- God's kingship over His people ended, and Israel, of its own accord, chose to be like the gentile nations. The aim of the covenant to create a model nation based on strict monotheism and obedience to the moral law to redeem the gentiles was not fulfilled.

- The heart-desires of sinful men like King David and his subjects navigated the destiny of Israel towards destruction rather than the wisdom and proper judgement of a caring God.

Chapter 4
A troublesome history – a Messianic hero who never came

After Israel expressed its desire to be like other nations and God's kingship of His people ended, Israel's gruesome history in the gentile world began. This period in Israel's history has been marked by internal and external conflicts, military defeats and expulsions, and the return and restoration of the Temple and worship in Jerusalem. In captivity, they longed to return to their homeland. When they returned, they rebuilt the Temple and resumed worship. Whenever Israel's security was in jeopardy, Messianic expectations took hold of their imaginations. Israel's history was in constant conflict with the gentiles, and Israel never became like other nations. The Temple in Jerusalem was destroyed twice, and Israel was forced into exile.

The end of God's engagement with Israel – Some years after the first Temple was built, Jerusalem was governed by corrupt kings and idolatry and sin were rampant among God's people. One King offered his loyalty to the King of Babylon. The King did evil in the eyes of Jehovah his God and rebelled against the King of Babylon to whom he offered his allegiance. Also, all the heads of the priests, and the people were continually acting treacherously according to all the abominable idols of the nations; and they defiled the house of Jehovah that He had sanctified in Jerusalem. And Jehovah the God of their fathers, sent to them by the hand of his messengers, rising early and sending, for He had pity on His people and on His dwelling place; but they mocked the messengers of God, despised His words, and scoffed at His prophets until the wrath of Jehovah went up against His people, until there was no healing.

(2 Chronicles 36:11-17) [3] The prophetic warnings to Judah and its rulers of the catastrophe that was to fall upon the nation because of their idolatry and sins were frequent and loud. God rejects Israel. "Behold, I will bring a nation on you from far way. O house of Israel, declares Jehovah. It is an enduring nation; It is an ancient nation, a nation whose language you do not know, nor understand what they say. Their quiver is as an open grave; they are all mighty men. And they will eat up your harvest and your food; and they will eat up your sons and your daughters; they shall eat up your flocks and your herds; they shall eat up your vines and your fig trees. One shall beat down your fortified cities with the sword, those in which you trust.

Yet even in those days declares Jehovah, I will not make a full end with you." (Jeremiah 5:15-18) [3] "So hear, O nations; and know, O congregation, that which is coming on them. Hear O earth; behold, I will bring evil on this people, the fruit of their thoughts. For they have not listened to My words, and My law, they also rejected it. Why is to Me, frankincense coming from Sheba, and the good cane from a far land? Your burnt offerings are not for acceptance, nor are your sacrifices sweet to Me." (Jeremiah 6:18-19) [3] Jehovah asked His servant to say, "Hear the words of the covenant, and speak to the man of Judah and to the inhabitants of Jerusalem; and say to them, So says Jehovah, the God of Israel, "Cursed is the man who does not obey the words of the covenant, which I commanded your fathers in the day I brought them out the land of Egypt, from the iron furnace, saying, Obey My voice, and do them accordingly to all that I command you, so that you shall be My people, and I will be your God." (Jeremiah 11:1-1-40) [3]

So, says the Lord Jehovah; "Strike with your hand, and stamp with your foot, and say, Alas, for all the evil abominations of the house of Israel for they shall fall by the sword, by the famine, and by the plague. He who is far off shall die by the plague; and who is near shall fall by the sword; and he who remains and is besieged shall die by the famine. So, I will fulfil My fury on them. And you shall know that I am Jehovah when their slain shall be in the midst of their idols all around their altars. On every high hill, in all the tops of the mountains, and under every green tree, and under every leafy oak; the place where they offered there a soothing aroma to all their idols. And I will stretch out My hand on them and make the land a desolation, even more desolate than the desert toward Diblath, in all their dwelling places. And they shall know that I am Jehovah." (Ezekiel 6:11-14) [3] "Make the chain; for the land is full of bloody judgements, and the city is full of violence. And I will bring the evil of the nations, and they shall possess their houses. And I will make cease the pomp of the strong ones; and their holy places shall be defiled. Anguish comes. And they shall seek peace, but none shall be. Disaster on disaster shall come, and rumour to rumour shall be. And they shall seek a vision from the prophet, but the law shall perish from the priest, and counsel from the elders. The king shall mourn, and the ruler shall be clothed with despair. And the hands of the people of the land shall be terrified. According to their way, I will do them; and according to their judgements. I will judge them. And they shall know that I am Jehovah." (Ezekiel 7:23- 27) [3] God's engagement with Israel as a nation ended well before the Babylonian invasion, and Israel stayed on a trajectory for more upheaval and conflicts in its history. Still, God continued to be loyal and responsive to those Israelites who were faithful and obedient to His moral law and called on His name in times of adversity and danger. Loyalty, faithfulness, good behaviour, sincerity of heart, reverence for and reliance on God resulted in God's favour and protection. Faith in God was personalised, and God was no longer responsible for the future affairs of Israel as a nation. There was a realisation of the need for inner renewal of the heart and spirit and the personal responsibility of the individual for his/her sin.

In 586 B.C., the kingdom of Judah was invaded, Jerusalem and the Temple were destroyed, and many people were taken into exile to Babylonia. In this period of captivity and enslavement, the Israelites longed to return to Jerusalem and rebuild their Temple. Recall that a temple was not mentioned in the original covenant, and God never wanted it because he preferred a tent. But, there were numerous prophecies about God's desire for His people to return to the holy land, rebuild the Temple, and resume worship and blood sacrifices. For example, So, says Jehovah, God of Israel, saying, "Write for yourself all the words that I have spoken to you in a book. For, lo, the days come, says Jehovah, that I will turn the captivity of My people Israel and Judah, says Jehovah; and I will cause them to return to the land that I gave to their fathers, and they shall possess it." (Jeremiah 30:1-3) [3] It is doubtful that these verses had divine origins when they are examined against the ones given in 2 Chronicles 36, Jeremiah 11, and Ezekiel 6&7. The books of Jeremiah, Lamentations, and Daniel contain prophecies, poems, and stories about the time of Israel in exile and their longing to return to the holy land.

After the exiles returned, they rebuilt the Temple and restored life, worship, and blood sacrifices. From the onset, there was severe opposition to the return of the Jews and the rebuilding of the Temple. The people living in the land considered Jerusalem an evil and rebellious city and found it difficult to cohabit the land with the Israelites. (Ezra 4:11-16) [3] Nevertheless, the Temple was rebuilt and dedicated, and the Israelites again established their presence in the holy land. Israel's history then took a turn for the worse.

A messianic hero who never came – Sometime after the second Temple was built in Jerusalem, Judaea became a part of the Ptolemaic Kingdom and Seleucid Empire around 200 BCE and was governed by the Hasmonean dynasty. The Maccabean dynasty is often used as a replacement for the entire Hasmonean dynasty. Judaea was heavily Hellenized, resulting in a struggle between Judaism and Hellenism and a revolt. There are various explanations for the revolt, ranging from religious to social and economic. Some scholars argue that the uprising started as a religious rebellion and gradually became a national liberation war. In 166 BCE, Judah Maccabee led an army to victory over the Seleucid dynasty, destroyed Hellenizing Jews and pagan alters in the villages, and forced Jews into outlawry. After victory was secured, the Maccabees captured Jerusalem, cleansed the Temple, and re-established Jewish worship. Jonathan Maccabee, the brother of Judah Maccabee, was installed as a high priest, and the Temple was re-dedicated. Judaea remained under the control of the Seleucid Empire until 129 BCE. The Hasmoneans ruled it until 63 BCE, when it became a client kingdom of Rome. In 37 BCE, the Hasmonean Kingdom was transformed into the Herodian Kingdom. This period of martyrs and heroic military leaders saved Judaism from annihilation and Israel from its enemies. [4] It further reinforced the notion of a Messianic saviour in the shape of a heroic military leader who would save Israel from its enemies during calamity. The Messianic expectations were grounded in the earlier period when God directed Israel's military campaign to confiscate land to house the twelve tribes of Israel. They were to be tested again when Rome occupied Judaea and the surrounding areas, imposing a harsh rule over their inhabitants.

According to some estimates, Jesus was born in Bethlehem in Judaea between 6 and 4 BC. John baptised him in the Jordan River. After having been baptised, the Spirit of God came down as a dove upon him, and behold, a voice out of heaven said, "This is My son, the Beloved, in whom I have found delight." (Matt 3:13-17) [3] Jesus started teaching in the synagogues, preaching the Good News about the Kingdom, and healing all kinds of diseases and sickness all over Galilee. (Matt 4:23-25) [3] Two events in Jesus' life had a profound impact on humanity. The Sermon on the Mount and his visit to the Temple in Jerusalem. In his Sermon on the Mount, Jesus said: Those who are spiritually poor will inherit the Kingdom of God; those who mourn will be comforted; those who humble themselves will receive what has been promised; those whose greatest desire is to do what God requires will be satisfied; those who are merciful to others will receive mercy; those who are pure in the heart will see God; those who work for peace will be God's children; and those who are persecuted because they do what God requires will inherit the Kingdom of heaven. (Matt 5:3-10) [3] Jesus entered the Temple of God and threw out those selling and buying in the Temple. And he overthrew the tables of the money changers and the seats of those selling the doves. And he said to them, it has been written, "My house shall be called a house of prayer, but you have made it a den of robbers." (Matt 21:12-13) [3] Jesus gave two great commandments. "You shall love the Lord your God with all your heart, and with all your soul, and with all your mind." "You shall love your neighbour as your-self." On these two commandments, all the Law and the Prophets hang. (Matt 22:37-40) [3] Jesus wept over Jerusalem. Jerusalem, Jerusalem, the one killing the prophets and stoning those sent to her. "How often I desired to gather your children in the way a bird gathers her chicks from under her wings. And you did not desire it. Behold, your house is left to you desolate. For I say to you, In no way shall you see Me from now on until you say, Blessed is He who comes in the name of the Lord." (Matt 23: 37-39) [3] Then Jesus spoke of the destruction of the Temple. And going out, Jesus left the Temple. And His disciples came to show Him the buildings of the Temple. But Jesus said to them, "Do you not see all these things? Truly I say to you, There will not at all be left one stone on a stone which in no way will not be thrown down." (Matt 24:1-2) [3] Jesus's prophetic words were fulfilled some years after his resurrection, and the Jewish presence in the holy land as a nation ended in catastrophe.

Jesus and his disciples gathered together some time before his crucifixion. Jesus taking a loaf, blessing it, He broke and gave to them. And he said, "Take eat, this is My body". And taking the cup, giving thanks, He gave to them. And they all drank out of it. And he said to them, "This is My blood, that of the New Covenant, which is poured out concerning many." (Mark 14:22-24) [3] The body of Christ, a new abode for Israel, symbolises the elect who are called to serve God, and the blood of Christ signifies his redeeming power.

In the original covenant, ancient Israel was commanded to be strict monotheists and obey God's moral law. In Jesus, these requirements were met, and he became a holy body in whom God resided, revealing Himself and healing many. In the new covenant, Israel is the body of Christ through which the elect are redeemed, thereby creating a model nation, as was the original intention of the old covenant. Jesus taught many parables, some regarding God, himself, humans, and the world, which are revealing. Here are some of Jesus' quotes: " Listen, a sower came forth, took a handful, and cast. Now, some fell upon the path, and the birds came and picked them out. Others fell upon rock, and they did not take root in the soil, and did not send up ears. And others fell upon the thorns, and they choked the seed; and the grubs devoured them. And others fell upon good soil, and it sent up good crops and yielded sixty per measure and a hundred and twenty per measure." (Thomas 9) [1] "I shall give you what eyes have not seen, what ears have not heard, what hands have not touched, what has not come upon the human heart." (Thomas 17) [1] "I shall choose you – one out of a thousand and two out of ten thousand. And they will stand at rest by being one and the same." (Thomas 23) [1] His disciples said, "Show us the place where you are, for we must seek it." He said to them, "Whoever has ears should listen. There is light existing within a person of light. And it enlightens the whole world: if it does not enlighten, that person is darkness." (Thomas 24) [1] " Whoever has become acquainted with the world has found a corpse, and the world is not worthy of the one who has found the corpse." (Thomas 56) [1] " Whoever drinks from my mouth will become like me; I, too, will become that person, and to that person the obscure things will be shown forth."(Thomas 108) [1]

In Rome, the tie of citizenship to the Roman commonwealth was highly desirable. But many in Judaea were excluded; this was discrimination against the Jews. The Jewish population demonstrated, and a riot broke out. The Roman governor responded with extreme force and sent the army. Many Jews were killed, and their properties were confiscated. In a court hearing before the emperor, the officials responsible were acquitted. This injustice infuriated the Jews. Religion was another source of contention with the Romans. In 38 A.D., the Romans marched on Jerusalem and erected cult statues of the emperor in the city and in the Temple enclosure to promote emperor worship. Such a display was contrary to the Jewish Torah, which is central to Judaism. In Jerusalem and many parts of Judaea, protestors gathered. Rome ordered objectors to be executed and the rest enslaved. The protests subsided when the Roman emperor died and was replaced with a more consolatory one. For most ordinary Jews, the burden of Roman taxes and other financial exactions was hard to bear. As the Roman emperors faced financial difficulties, they raised taxes, which hit Judaea very hard. The emperor sent his soldiers to Jerusalem, forcing their way into the holiest of places, scattering the sacred objects, and violently dispersing the priests and protestors who stood in the way and took the money. More riots escalated, houses were plundered, thousands were killed, and the instigators of the riot were crucified as a lesson to others. There were more riots and significant disturbances, and as violence and bloodshed continued, the authority of the local leaders and priests collapsed, and the nationalists advocated armed resistance. [5]

The nationalists organised attacks on the Roman soldiers and pushed them back. The Roman authorities tried to bring peace by negotiating with the nationalist leaders, but they were unsuccessful, and the slaughter of the Roman soldiers by the outraged Jews continued. It looked as though the Roman grip on Judaea was weakening. The Romans brought in more military units but were repeatedly beaten back after suffering heavy casualties. The Jews in Judaea and other provinces were overjoyed and considered their victory a miracle. They thought

God played His part in defeating the almighty power of Rome. However, it was highly unlucky that God had any interest in the fate of Israel, and the coming events would demonstrate it beyond a doubt. The high priests decided that Judaea must fight Rome. The people agreed and appointed some priests to lead the war strategy. The high priests and the people prepared for war. The Romans were concerned that the uprising of the Jews in Judaea could have triggered revolts in other towns and provinces where Jews lived, so they prepared for war. The Roman general Titus led his armies and broke through the Jewish defences, killing thousands. Most of the remaining rebels were quickly and easily winkled out and overpowered, and the Romans took control of the town. [5]

Internal fighting among the Jews and mass-suicide pacts were a common sight. There were frequent invasions of the Jewish towns and provinces, where thousands were slaughtered, and many more were sent into slavery. In the final act of betrayal, the nationalist and moderate Jews killed each other, and the Temple complex became a battleground. In March 70 AD, Jerusalem and the most sanctified part of the Temple, the iconic epicentre of the Jewish faith, were looted and left to burn. A massive crowd of civilian men, women, and children had gathered in the Temple area, believing they would find signs of deliverance from God, but it never happened. Pagan standards were brought into the Temple complex, erected, and sacrifices were offered to the emperor. The Romans finally destroyed the Temple as Jesus prophesied. The Jews were expelled from the land of Judaea and the surrounding towns and villages to start two thousand years of persecution, diaspora, pogroms, displacement, expulsion, and loss of property, leading to the Holocaust in the 20th century. [5] In 1946 AD, Jews returned to the holy land to establish their state, Israel.

The Messianic expectations of Israel failed to deliver a heroic military leader like Judah Maccabee to save Judaea, the Temple, and its priests and people from the Romans. Jesus Christ could never have been accepted as a Messiah because the historical structure of Judaism would only permit a heroic military leader to deliver Israel from its mortal enemies like Rome. The Messianic mission of Jesus was for a relatively small number of the Israelites who remained loyal and obeyed God's moral law. The early

Christian converts were mostly Jewish, but they were too few to transform the entire Judaism into a Christocentric faith. Eventually, the Christian message found its way to the gentiles, and a worldwide movement emerged, which persecuted the Jewish people for centuries.

The origin of Israel's demise lies in three events in its history: when the Israelites, despite repeated warnings, rejected God as their king and expressed their desire to become like other nations (1 Samuel 8&12); when they built a temple to worship God and offered blood sacrifices regardless of God's displeasure (1 Chronicles 17); when God disowned them because of their frequent idolatry and immoral behaviour (Jeremiah 11 & Ezekiel 7). Consequently, their destiny as a nation was no longer in God's hands. It is incomprehensible that God could have continued to interact with a nation that no longer saw Him as its source of strength, guidance, wisdom, and faith and instead chose to be like other nations, relying on mortal and frequently corrupt kings. The original covenant demanded strict monotheism and obedience to God's moral law, and there were no Messianic promises, temple-centred worship, or unwarranted blood sacrifices.

But as time passed, the Israelites demanded favours that were not in line with the covenant's terms, which their ancestors had agreed to, nullifying the covenant. It is a mystery why God willingly agreed to meet their demands if all the signs pointed to the destruction of the covenant. It's as if God became a service provider for His people rather than the other way around. At times, it seemed as if Israel had a covenant with God. In response, God provided what the Israelites demanded, and the Israelites, in return, promised to abandon idolatry and improper conduct, only to revert to their bad habits again and again. It could be argued that Jewish history, with all its catastrophes, crises, and bloodshed, is the unintended consequence of the failure of the original covenant with ancient Israel, despite God's repeated attempts to keep His People on the right track.

Summary – The prophetic warnings became more frequent as the Israelites drifted away from the covenant God made with their ancestors. The original covenant contained no Messianic promises or mention of temple-based worship or arbitrary blood sacrifices. But, the Israelites were slowly and surely adapting to both, and their national destiny diverged from God's plan for them. As they settled in the holy land, they were often ruled by gentile empires and struggled for freedom and independence, relying on military heroes to save them from their enemies. Temple worship and blood sacrifices became central to their religious beliefs and practices. The Temple in Jerusalem became the epicentre of the Jewish national identity and existence and came under attack over centuries of conflict and conquest. The Romans finally destroyed it, and the Jewish presence in the holy land ended. No Messianic hero came to rescue Israel when it needed it most. This was followed by two millennia of intense Christian persecution, which resulted in the diaspora, expulsions, confiscation of property, and pogroms. [6] Some Jews returned to the holy land in 1946 to create their own country, where they are now secure and independent. It looks as if the history of Israel has gone around in a circle: the invasion of Judaea; the destruction of the Temple; diaspora and slavery in foreign lands; a yearning to return and rebuild the Temple; and once returned to the holy Land, an invasion occurred again. Although the state of Israel is secure for now, an essential piece of the jigsaw of Israel's destiny is missing. The missing piece will be discussed in the next chapter.

Conclusions

- The Israelites' parting from the original covenant God made with their ancestors set them on a course, leading to unintended Messianic expectations, temple-centred worship, and unauthorised blood sacrifices. The consequences became painfully clear as the Israelites' history unfolded.

- The unintended Messianic expectations led to bitter disappointment for Israel during adversity and national emergencies. The building of a temple in Jerusalem made the Israelites a target for conquest by gentile nations. Moreover, it corrupted their original faith and traditions after exposure to pagan ideas, intermarriages, and false religions when they were in exile.

Chapter 5
The missing piece of the jigsaw – the coming of the two Messiahs

The history of the Israelites over the centuries has been gruesome, and it ended seventy-five years ago when the state of Israel was established. Since its creation, the country has experienced internal unrest and military invasions by its neighbours but has managed to survive. It almost looks as if Israel is like other nations. But this is grossly misleading. As past events have shown, Messianic expectations have been an intrinsic part of the national existence of Israel. Jewish history dictates that Messianic expectations should be revived. World events are heading towards a major military confrontation and global devastation. The Pagan Roman Empire will rise from the ashes of the coming conflict, led by a heroic political and military leader, a Messianic figure who will order the construction of a temple in Jerusalem where he will establish himself as God. God's risen Messiah and the Jewish Messiah will come face-to-face at this unique time in history.

The birth of the state of Israel – After the destruction of the second Temple in Jerusalem by the Romans and the death of thousands of people, the remaining Jewish population of Judaea and the surrounding villages and towns were expelled into exile. In the diaspora, they suffered from intense persecution and pogroms and longed to return to the holy land. Some important geo-political events took place, which made their return possible. The Zionist movement, the Balfour Declaration, and the Holocaust in Europe are important events in modern Jewish history.

Antisemitism, the Zionist movement, and the Balfour Declaration – For centuries, antisemitism has been based on the belief that Jews were responsible for killing Jesus. Even after the Reformation in the 16th century, Christians continued persecuting Jews and subjecting them to pogroms and expulsions. The second half of the 19th century witnessed the emergence of pseudo-scientific racism, which considered Jews a race whose members were engaged in mortal combat with the Aryan race for world domination. So, humans were no longer considered racial equals with equal hereditary value. And new racism coined phrases like " the Jews are our misfortune", which reinforced deeper antisemitism among Europeans and was used by Nazi ideologists. Religious antisemitism, or anti-Judaism, is hostility towards Jews because of their perceived religious beliefs. It is proposed that antisemitism and attacks against Jews would stop if Jews stopped practicing Judaism or changed their public faith by conversion to the official or right religion, meaning Christianity. Even after conversion, hostility towards Jews continued. [7]

The Zionist movement emerged in the late 19th century in Central and Eastern Europe. It championed the creation and support of a homeland for the Jewish people in the area known as Palestine and Canaan, or the holy land. This was in response to centuries of antisemitism and persecution. [8] The region known as Palestine was captured by the Ottoman Empire in 1516 A.D. until Egypt took control in 1832 A.D. The area was returned to the Ottoman Empire some eight years later. After the end of the First World War, the Ottoman Empire lost control of Palestine to the British Empire, which administered it between 1920 and 1948. This period is referred to as the "British Mandate." The Balfour Declaration in 1917, during the First World War, supported a national homeland for the Jewish people in Palestine. [9]

There is a variety of Zionism called "cultural Zionism." Among them, religious Zionists supported Jews upholding their Jewish identity by adhering to their religious tradition, and political Zionism, championed by Theodor Herzl, was very influential. The political Zionists encouraged Jewish migration to Palestine and recruited European Jews to immigrate there. The Jews living in the Russian Empire experienced raging antisemitism too. The sponsors of Zionism consider it a liberation movement to repatriate persecuted people to their ancestral homeland. Since the Jewish state was established in 1946, Zionism has struggled against the continuous threat to its existence and security from its enemies. Theodor Herzl considered antisemitism to be an eternal feature of all societies in which the Jews lived as minorities, and only self-government could allow Jews to escape persecution. [8] The state of Israel was formally founded in 1946 after thousands of Jews immigrated to Palestine from European countries during the Holocaust, and soon Israel found itself engaged in a war of independence with its Arab neighbours.

The state of Israel - Soon after the state of Israel was recognised, the land of Palestine, which was still under the British Mandate, was partitioned by the United Nations resolution passed in 1947. The resolution recommended the creation of independent Arab and Jewish states and a Special International Regime for the city of Jerusalem. The Arab Palestinian leadership refused to accept the resolution. The Arab League refused to accept the UN partition plan and proclaimed the right of self-determination of the Arabs across the whole of Palestine. [12] In 1948, Jewish immigrants, veterans, and Holocaust survivors, mainly from World War II, began arriving in Israel in large numbers. The population soon swelled to make Israel and its military forces more formidable. The Arab armies invaded Palestine, starting the Arab-Israeli wars. After some early military setbacks, Israel received more weapons, mostly from Western countries, and the tide of war changed in its favour. The Arab armies suffered defeat after defeat, and peace treaties were signed with Arab nations, bringing all hostilities to an end. Since the peace treaty was signed, Israel has enjoyed secure borders, but there is internal unrest among its Palestinian population. Although Israel is now an established country and internationally recognised, there are concerns about its conduct and treatment of other non-Jewish minorities, like the indigenous Palestinians. In 1975, the United Nations General Assembly passed a resolution designating Zionism as "a form of racism and racial discrimination." The resolution was eventually revoked in 1991. Israel's borders are secure, and it has achieved international status with the full support of Western nations, mainly the English-speaking people. [13] The question is, "What is Israel's future?" The answer lies in its history.

When Israel's history is examined retrospectively, a pattern emerges. As mentioned in Chapter 4, the Israelites' departure from God's original covenant with their ancestors led them to unintended Messianic expectations, temple-centred worship, and blood sacrifices. Unintended Messianic expectations led to bitter disappointment when they failed to prevent the Babylonian invasion, the Roman destruction of Jerusalem's second Temple, and the exile of its inhabitants to foreign lands. The building of a temple in Jerusalem made the Israelites a target for conquest by gentile nations. The logic of Israel's history dictates that the Messianic expectations must be revived when the Jewish people live as a nation in the holy land, as they did before the Babylonian invasion and the Roman destruction of

Judaea. But will this happen in Israel? The apostle Paul prayed that his own people, the Jews, would be saved. (Romans 10:2) Today, an unprecedented event is taking place in Israel: Jews are finding Jesus as their personal Saviour and converting to Christianity. It remains to be seen whether the converts will constitute a big force in Israel's political and religious landscape or not. There is also significant movement on the part of the orthodox Jews, who believe in a future Messiah and the re-building of the Third Temple in Jerusalem. It is not yet clear how the events will unfold, but if the orthodox Jews were successful in rebuilding the Temple, then one may expect major upheaval to occur in the holy land, as stated in the Bible, which will affect us all. "Do not let anyone deceive you in any way, because that day will not come unless first comes the falling away, and the man of sin is revealed, the son of perdition, the one opposing and exalting himself over everything being called God, or object of worship. So as for him to sit in the temple of God as God, showing himself that he is a god." (2 Thessalonians 2:3-4) [3] A temple in Jerusalem must be built for this Messianic figure to fulfil his historic mission as described in the biblical prophecies. As stated earlier, there are at least two powerful factions in Israel today: religious Zionists and political Zionists. [8]

The Temple Movement began in Israel in 1987, and is preparing for the rebuilding of the Third Temple in Jerusalem. This movement is spearheaded by Orthodox Jews who are reviving the Sanhedrin, the religious body that supervises the legal issues related to the administrative duties of the Temple. The Jewish leaders in this movement believed that the Jewish people were not living the spiritual lives God intended because of the absence of the Divine presence in the world. It is believed that the building of the Temple plays a role in the world's redemption, and it will happen only when the Temple is built. But many religious Jews do not support the rebuilding of the Third Temple. Nevertheless, priests are being trained, and ritual vessels needed for the function of the Temple are being produced. The mission statement says, "For Orthodox Jews committed to re-establishing the Temple, both the present problems of the world and the problems faced by the Jewish people will be solved only by rebuilding." [14] A red heifer must be sacrificed before the Temple can be rebuilt, and preparations are being made in Israel to breed and sacrifice a red heifer in line with the biblical tradition.

And Jehovah spoke to Moses and to Aron, saying, This is the statute of the law which Jehovah has commanded, saying, "Speak to the sons of Israel, that they bring you a red heifer, a perfect one, in which there is no blemish, on which no yoke ever came. And you shall give her to Eleazar the priest, and she shall be brought forth outside the camp; and she shall be slaughtered before his face. And Eleazar the priest shall take of her blood with his finger and shall sprinkle of her blood toward the front of the tabernacle of the congregation seven times. And the heifer shall be burned before his eyes; her skin and her flesh, and her blood with her dung, shall be burned." (Numbers 19: 1-5) [3] If indeed the red heifer's sacrifice was commanded by God to Eleazar the priest, when the tabernacle of the congregation was in place, then God was the one who chose the time and the reason for the sacrifice, and the priest who carried it out. There is no reason to believe that sacrificing a red heifer in a temple that God will not dwell in will have any religious significance at all in an age when the body of Christ is the focus of the new covenant. Recall when the Israelites removed the Ark of the Covenant of Jehovah and took it into battle, expecting it to save them from defeat at the hands of their enemies, and it did not? (1Samuel 4:1-4) [3] The Israelites never benefited from religious relics, ceremonies, or rituals in the past, and it is not clear if they will do so now or in the future. The rituals of blood sacrifice in Judaism that are not authorised by God are highly questionable and may have pagan origins. Jesus did not sacrifice blood. For example, no blood was shed to forgive sins. "Your sins are forgiven" was the cry of Jesus. In any case, the original covenant ended millennia ago, years before the Babylonian invasion, and the heifer's sacrifice belongs to a time when God had full engagement with collective Israel as a nation and He protected His people. All the signs indicate that the world is rapidly moving towards the rebuilding of a third temple in Jerusalem and that the Jewish Messiah will appear on the scene soon. For these events to happen, a falling away and a major global conflict must engulf the world. "See that not any leads you astray. For many will come in My name, saying, I am the Christ. And they will cause many to err. But you are going to hear of wars and rumours of wars. See, do not be disturbed. For all things must take place, but the end is not yet. For nations will be raised against nations, and kingdom against kingdom; and there will be famines and plagues and earthquakes against many places. But all these are the beginning of throes." (Matt 24:4-8) [3]

From the ashes of a global conflict and mayhem, the Pagan Roman Empire will rise. "And I stood on the sand of the sea. And I saw a beast coming up out of the sea, having seven heads and ten horns, and on his horns ten diadems (crowns), and on its heads, names of blasphemy. And the beast which I saw was like a leopard, and its feet as of a bear, and its mouth as a lion's mouth. And the dragon gave its power to it, and its throne, and great authority. And I saw one of his heads, as having been slain to death, and its deadly wound was healed. And all the earth marvelled at after the beast." (Revelation 13:1-3) [3] These verses describe the events leading to the rise of the beast, or the Pagan Roman Empire, on the global stage. This new world order will be headed by a figure (probably a heroic political and military leader fulfilling the Jewish Messianic expectations) who is described in the following verses: "And they worshipped the dragon who gave authority to the beast, and they worshipped the beast, saying: Who is like the beast, who is able to make war with it? And a mouth speaking great things and blasphemies was given to it. And authority to act for forty-two months was given to it. And it opened his mouth in blasphemy toward God, to blaspheme His name and His tabernacle, and those tabernacling in heaven." (Revelation 13:4-6) [3] When the temple referred to in 2 Thessalonians 2:3-4 is built and the expected Jewish Messiah appears, at this unique moment in human history, God's Messiah, Jesus Christ, will return to face the man of sin and destroy him forever, thus fulfilling God's providence and plan for the coming of His Kingdom and the redemption of humankind. This troublesome Jewish history and the gentile age will end.

Summary – The history of Israel since the Babylonian invasion has been a time of extreme trouble, violence, and suffering, culminating in the birth of the Zionist movement, the horrors of the Holocaust, and the creation of the state of Israel. Since its creation, Israel has matured as a nation with secure borders with its neighbours and is now a major military power in the Middle East. The history of Israel has been based on two facts: temple-centred worship and Messianic expectations, neither of which were in the original covenant God made with their ancestors. When the Messiah (the man of sin) appears, the construction of a temple will begin. But for the Messiah to emerge on the global scene, the beast must rise first. We are now witnessing major events taking place in the world that will soon lead to the greatest upheaval humankind has ever experienced in its long history. Then Jesus said, "Then when you see the abomination of desolation which was spoken of by Daniel the prophet, standing in the holy place—the one reading, let him understand—then let those in Judaea flee on the mountains; the one on the housetop, let him not come down to take anything out of his house; and the one in the field, let him not turn back to take his garment." (Matt 24: 15-18) [3] So much horror is to come yet, but for those who put their trust in God, Jesus had this to say, " Do not let your heart be troubled; you believe in God, believe also in Me." (John 14:1) [3]

The prophecies regarding Gog and Magog – Gog and Magog are entities which God despises because of their hostility to Israel. "And you, son of man, prophesy against Gog and say, So, says the Lord Jehovah: Behold, I am against you, o Gog, the Prince of Rosh, Meshech, and Tubal. And I will turn you back and lead you on. And I will bring you up from the recesses of the north and will bring you on the mountains of Israel. And I will strike your bow out of your left hand, and I will cause your arrows to fall out of your right hand. You shall fall on the mountains of Israel, you and all your bands, and the people who are with you. I will give you for food to the birds of prey, a bird of every wing; and to the beast of the field. You shall fall on the face of the field, for I have spoken, declares the Lord Jehovah. And I will send a fire on Magog and on the secure inhabitants of the coasts. And they shall know that I am Jehovah." (Ezekiel 39:1-6) [3] There are also prophecies in the books of Revelation regarding Gog and Magog. "And whenever the thousand years are ended, Satan will be set loose out of his prison, and he will go to mislead the nations in the four corners of the earth – Gog and Magog – to assemble them in war, whose number is as the sand of the sea. And they went up over the breath of the land and encircled the camp of the saints, and the beloved city. " (Revelation 20:7-9) [3] It is widely believed that Gog and Magog will be a military alliance against the state of Israel in the Middle East. This alliance will attack Israel and be defeated by Divine intervention; i.e., the long-awaited Jewish Messiah who will appear to save Israel. It cannot be denied that major military conflicts will take place in the future in the Middle East involving the state of Israel, as they have in the past. But this interpretation of the verses in Ezekiel 39 may be misleading. God's Kingship of Israel as a nation ended millennia ago (1 Samuel 8:7-9) [3] and the New Covenant by Christ replaced the old one, bringing God's plan to transform ancient Israel into a model nation to an end. (Luke 22:19-20) [3] It is worth remembering the following prophecy: " Because we are not wrestling against flesh and blood, but against the rulers, against the authorities, against the rulers of this world, of the darkness of this age, against the spiritual powers of evil in the heavenlies." (Ephesians 6:12) [3] Hence, since the New Covenant has made the Body of Christ the dwelling of Israel where redemption of the elect is taking place, the verses in Revelation 20 are more relevant and suggest that Israel, the true Church, has been under attack for some time by the forces of evil and their principalities, Gog and Magog. Western Christian culture and heritage are in peril. Idolatry, immorality, inequality, ungodliness, lawlessness, crime, corruption, injustice, and false religious teachings are wide-spread in the Western world, which traditionally held strong Christian ethical beliefs and values. The exact identity of Gog and Magog may not be readily known, but it is likely that they are of a spiritually demonic nature, and their destructive powers and influence are everywhere to be seen. It is only through faith and obedience to God's moral law that humans can overcome this evil and escape the judgement which is coming on this world. "For the rest, my brothers, be made powerful in the Lord and in the might of His strength." (Ephesians 6:10) [3]

Summary - World events are slowly but surely leading to the rise of the Pagan Roman Empire, or the Beast mentioned in the books of Revelation. When the New World Order is put into place, the Wicked One will emerge on the scene, claiming to be the Messiah, leading humanity to the greatest deception. The false Messiah will then demand the building of a temple in Jerusalem, where he will sit and blaspheme the name of God. It will be at this historic moment that God's anointed King, Jesus Christ, will come to defeat and destroy falsehood and usher in a new age, the Kingdom of God. The golden Jerusalem will come down from the clouds, and the Holy God will take His place, uniting with the redeemed in Christ. This is when eternity will begin.

Conclusions

A chronology of the events that may be coming soon -

- The world will be engulfed in a major global security crisis when nations rise against nations, leading to wars and unimaginable devastation.

- The Pagan Roman Empire (the Beast) will rise from the ashes of the coming global conflict headed by a Messianic figure.

- The false Messiah will order the building of the Third Temple in Jerusalem; the restoration of the organisation of the traditional religious Judaic rituals, and the blood sacrifices.

- The false Messiah will appear in the Temple, claiming to be God, thereby making the Temple an abomination to the Holy God.

- God's anointed King, Jesus Christ, the risen Messiah, returns and wages war against the abomination in the Temple, destroying it forever.

- The Kingdom of God arrives; the rightful King takes to his throne; God resides in the Golden Jerusalem; and the redemption of the elect begins.

Chapter 6
Summary and conclusions

The Bible and the Jewish scriptures contain wisdom that has no rival. The Bible contains God's revelations and a rich history of God's interactions with His people, as well as prophecies and warnings. No source of human knowledge will ever match that of the Bible. However, the Bible contains parables, symbols, and metaphors that must be deciphered to understand the biblical narrative. When information in the Bible is studied in conjunction with Jewish history, a troubling story emerges.

The religious soul of humankind – The religious experience of mankind has been and continues to be very colourful and wide-ranging in its nature and scope. Throughout history, men have believed in gods and worshipped them. Blood sacrifices in temples dedicated to gods and idols have been a widely practiced ritual to express devotion to gods for the fear of the unknown and unseen, adoration of their majesty and power, and to please them to secure their favours and blessings. For example, in Polynesian religion, the gods visit when dwellings are prepared for them and physical rituals such as human sacrifices or the recitation of magic words or sounds are performed to please them. Once this is done, the gods are sent back to where they came from. The Incas used temples for sacrifices, including human sacrifices and there were complex rituals performed related to calendars describing the movements of various heavenly bodies. In the Aztec tradition, the universe was very unstable, and needed continuous sacrifice to remain ordered and stable. This in turn perpetuated warfare to capture men for sacrifice to gods. In some extreme rituals, a victim's chest would be opened and the still beating heart torn out and presented to the gods. [15,16] Idolatry, temple-worship, and blood sacrifices have their origins in paganism and have found their way into the main stream of Abrahamic religions, corrupting them.

The Israelites were commanded by God to have faith in Him and to follow His moral laws. Monotheism and strict obedience to moral law have proven to be incompatible with human nature, frequently clashing with man's natural religious instincts to worship pagan gods and offer blood sacrifices to idols in temples dedicated to them. The history of the people of God has shown that some embraced faith in the Holy God and obeyed His moral law, while the majority did not. This, coupled with fear, superstition, and ignorance, brought them into conflict with the Holy God and prevented them from enjoying a unique relationship with the Divine.

The Israelites' frequent idolatry and immoral conduct indicated a malignancy in the soul of humankind (Jews and gentiles) that draws it to paganism, temple-worship, and blood sacrifices. It is as if humans have an inbuilt tendency or an instinct to commit these acts, and by doing so, they violate God's commandments and commit sinful acts, displeasing God. The biblical narrative is a graphic description of the Israelites' struggle with strict monotheism and God's moral law, leading them towards their natural instincts of idolatry and immoral conduct. They became like other nations that God wanted to redeem through the covenant with their ancestors. There is no evidence to suggest that the frequent warnings, pleadings, and punishments could have ever stopped this slow but inevitable transition in the lives of God's people.

The falling away - It is a mystery why God conceded so much by allowing the Israelites to have military leaders, kings, kingdoms, and a temple. These provisions were not in the original covenant with their ancestors and did not stop the Israelites from committing sins. Nevertheless, it was inevitable that some of God's people would fall away from the covenant sooner or later. In the diaspora, the Israelites were persecuted for centuries, and since they returned to their home land in the twentieth century, they have been in constant conflict with their neighbours. It is obvious that nothing less than redeeming the heart can ever restore the human relationship with God. Jesus Christ is the redeemer, and in his heart, God's moral law is written.

Was Jesus Christ the Messiah? – Jesus revealed himself to be the Son of Man; the bread of life; the vine; the light of the world; the good shepherd; the way, truth and life; the door to a green pasture; the resurrection and life; and a bridegroom. (Matt:9, Luke:18 and John:6,7,8,9,10,11&14) [3] To the scribes and the Pharisees, he said, " I am the Light of the world; he following Me will in no way walk in the darkness but will have the light of life. Then the Pharisees said to Him, you witnessed concerning yourself; your witness is not true." (John 8:12 13) [3] Jesus reveals himself to be the Messiah to a Samaritan. " The woman said to Him, I know that Messiah is coming, the one called Christ. When that

One comes, He will announce to us all things. Jesus said to her, I AM , the one speaking to you." (John 4:25&26) [3] Jesus said to Martha, " I am the Resurrection and the Life; the one believing into Me, though he die, he shall live. And everyone living and believing into Me shall die to the age. Do you believe this?" She said to Him, "Yes, Lord, I have believed that You are the Christ, the Son of God who comes into the world." (John 11:25-27) [3] Jesus was sent by God to preach the Good News about the Kingdom of God. And day coming, going out He went into desert place. And the crowds looked for Him and came up to Him, and held Him fast, not to pass away from them. But He said, to them, " It is right for Me to preach the gospel, the Kingdom of God, to the other cities, because I was sent on this mission." (Luke 4:42-43) [3]

What did Jesus say to his disciples? And coming into the parts of Caesarea of Philip, Jesus questioned His disciples, saying, " Whom do men say Me to be, the Son of man?" And they said, "Some say John the Baptist and others Elijah; and others Jeremiah, or one of the prophets." He said to them, "but you, whom do you say Me to be?" And answering, Simon Peter said, " You are the Christ, the Son of the living God." And answering Jesus said to him, " Blessed are you,

Simon, son of Jonah, for flesh and blood did not reveal it to you, but My Father in Heaven."(Matt 16:13-17) [3]

To the scribes and Pharisees who were ignorant of the truth, Jesus revealed himself as the light of the world; to the faithful and law-abiding Israelites as the Messiah; and to the rest of humankind as the bread of life, the vine, the light of the world, the good shepherd, the way, truth, and life; the door to green pasture; the resurrection and life; a bridegroom, and the Son of Man. To the righteous, Jesus is the hope of the kingdom to come, and to the unrighteous, he is the cause of their indignation and damnation.

Summary – Ignorance and spiritual blindness are intrinsic to the human soul. This is due to our natural instinct to worship pagan gods and sacrifice blood in temples dedicated to them. This was demonstrated graphically in the long history of ancient Israel. No demand for strict monotheism and the imposition of external moral law can ever remedy this malignancy; only full purification (redemption) in the person of Jesus Christ can. When the moral law is written on the human heart, as it was done with Christ, then humans will become God's children, as it has been planned since the foundation of the world.

Conclusions

- The biblical narrative suggests that humans' natural instinct to worship pagan gods and idols and sacrifices of blood in temples dedicated to them prevented most of the Israelites from remaining strictly monotheists and obeying God's moral law.

- Jesus Christ is a blinding light to the unrighteous and a Messiah, a guiding light, and a beacon of hope for the righteous.

"The righteous shall inherit the earth and live on it forever." (Psalm 37:29) [3]

The disciples said to Jesus, "When will you be shown forth to us and when shall we behold you?" Jesus said, " When you strip naked without being ashamed, and take your garments and put them under your feet like little children and tread upon them, then you will see the child of the living. And you will not be afraid." (Thomas 37) [1]

Abraham's faith

"My God, God most high, You alone are my God, and You and your dominion have I chosen. Deliver me from the hands of evil spirits who have dominion over the thoughts of men's hearts, and let them not lead me astray from You, my God." [2]

Take Jehovah as your Lord and follow His moral law, or love the world and walk through the valley of death as ancient Israel did.

As was the case with Abraham, this yearning comes from the heart. So, let this be the prayer of the righteous.

References

1. Bentley Layton, The gnostic scriptures. SCM Press Ltd, 1987   (ISBN 0-334-02022-0)

2. Joseph B. Lumpkin, The books of Enoch. Fifth Estate Publishers, Blountville, AL   35031. 2011 (ISBN: 9781936533077)

3. Jay P. Green, Sr. The interlinear Bible. Hendrickson Publishers, 1986   (ISBN 978-1-56563-977-5)

4. https://en.wikipedia.org/wiki/Maccabees. Date visited: 08-09-2022

5. Simon Baker, Ancient Rome. The Rise and fall of an empire, BBC Books, 2007   (ISBN 978-1-846-07284-0)

6. Ray Montgomery, Bob O'Dell. The List: Persecution of Jews by Christians throughout history. Root Source Press. Jerusalem, Israel 2019 (ISBN: 978-965-7738-13-9)

7. https://en.wikipedia.org/wiki/Antisemitism Date visited: 13-09-2022

8. https://en.wikipedia.org/wiki/Zionism Date visited: 13-09-2022

9. https://en.wikipedia.org/wiki/History_of_Palestine
   Date visited: 13-09-2022

10. https://en.wikipedia.org/wiki/Balfour_Declaration
    Date visited: 13-09-2022

11. https://en.wikipedia.org/wiki/History_of_Israel
    Date visited: 13-09-2022

12. https://en.wikipedia.org/wiki/United_Nations_Partition_
    Plan_for_Palestine   Date visited: 14-09-2022

13. Ali Ansarifar, The British and American Empires and the
    State of Israel. Kingdom Publishers, London 2022
    (ISBN: 978-1-913247-98-0)

14. https://www.jewishvoice.org/read/article/update-
    buildingthird- temple   Date visited: 14-09-2022

15. Ninian Smart, The World's Religions. Cambridge University
    Press. 1998   (ISBN 0-521-63139-4)

16. Ali Ansarifar, The March to the Armageddon. Balboa Press,
    2022   (ISBN 978-1-9822-8377-3) (sc)
    (ISBN 978-1-9822-8378-0) (e)

Epilogue

The Bible and the Jewish scriptures contain wisdom that has no rival. The Bible contains God's revelations and a rich history of God's interactions with His people, as well as prophecies and warnings. It contains parables, symbols, and metaphors that must be deciphered to understand the biblical narrative. When information from the Bible is studied in conjunction with Jewish history, an interesting fact emerges. One wonders why the people whose ancestors were so fortunate and privileged to meet the Holy God have such a dismal history. Their history is riddled with idolatry and disobedience to God's moral law, and this needed explaining.

This book demonstrates that a malignancy in the souls of the Israelites led them to idolatry and immorality, alienating them from God. When the Israelites rejected God as their King and expressed their desire to be like other nations, God's engagement with Israel as a nation ended. They then developed a religious tradition that was based on temple-worship and blood sacrifices, coupled with Messianic expectations that were not fulfilled.

In the modern age, humans alienate themselves from God through their desires for fame and fortune. Fame validates human selfishness and flattery, and fortunes validate human greed for material goods, which can be excessive and sometimes immorally obtained. And when individuals turn to God, seeking a saviour to alleviate their suffering and difficulties, their Messianic expectations will not be met. This is a universal problem with humankind and will continue to separate humanity from God. Redemption in the person of Jesus Christ is a radical solution to changing human hearts and saving humans from perishing.

Afterword

No source of human-generated knowledge or wisdom can or will explain the predicament that humankind has been in since its creation. The Bible provides a graphic description of the conflict between God's will and human nature. It is well documented that humans throughout their history have been religious and practiced their religiosity by worshipping pagan gods and idols, building temples, and sacrificing blood in their honour. God commanded strict monotheism and full obedience to His moral law. But this proved to be impossible to achieve by some of His people and brought retribution and punishment for disobeying them. The malevolence in human soul, which forced the Israelites into idolatry and sin, is universal, irrespective of whether one is a Jew or gentile. Unless there is a fundamental change in the hearts of humans, God's relationship with humankind will remain marred forever. By God's grace and mercy, the moral law was written on Jesus's heart to make him godly and moral, as God intends all humans to be.

About the author

Dr. Ali Ansarifar has been living in the U.K. for over 40 years. He was awarded a bachelor's degree, a doctorate in Materials Science from Queen Mary College, the University of London, and a Diploma in Interface Science from Imperial College, University of London. He worked as a postdoctoral research assistant at Imperial College, London, and Cavendish Laboratory, Department of Physics of the University of Cambridge. He was an upper senior research scientist in a rubber research and development centre in Hertfordshire, U.K., and a lecturer in Polymer Engineering in the Materials Department at Loughborough University until he retired as a senior lecturer. He has given lectures, seminars and workshops in the United States, the United Kingdom, Europe, the Middle East and Southeast Asia, published over 150 technical research papers in peer-reviewed international scientific journals and technical magazines for the polymer and tire industries and textbooks, and contributed chapters to scientific books. He has been on the editorial board of rubber and adhesion scientific journals and has been awarded prizes for his scientific publications. He is a Fellow of the Higher Education Academy U.K. and a servant of Jesus Christ.

Final remarks

For millennia, humankind has wondered about the wisdom in the Bible. The biblical revelation has inspired the curiosity, interest, faith, intellect, and hope of humans for generations. But, what does the biblical narrative convey? Opinions may vary. However, it is abundantly clear that it is not God's commands that make humans faithful and obedient to His moral law; rather, it is the inner hearts of humans that surge towards faith and divinity. One may say that godliness is humans' most noble asset, and that it comes forth naturally from the inner heart rather than being demanded externally by a supremely powerful deity. As precious as the Abrahamic faith is, alas, it is so rare among humans. The Old Testament in the Bible is a graphic description of the struggle between God and humans who rejected Him for lack of Abrahamic faith in their inner hearts. Humans have either got it in their inner hearts or they do not. This was a moment of truth in the entire history of human existence on earth.

Books by the author

1. Sharing the Faith. Kingdom Publishers, London 2021. ISBN: 978-1-913247-54-6

2. The Bible Story of Mankind. A Covenant with God with no Get-out Clause. Balboa Press, UK 2021. ISBN: 978-1-9822-8394-0 (sc), ISBN: 978-1-9822-8393-3 (e).

3. Why Did God Create Mankind? The Problem of Duality with God. Balboa Press, UK 2021. ISBN: 978-1-9822-8439-8 (sc), ISBN: 978-1-9822-8440-4 (e).

4. The British and American Empires and the State of Israel. Until the Kingdom of God Comes. Kingdom Publishers, London 2022. ISBN: 978-1-913247-98-0

5. The March to the Armageddon, Balboa Press, UK 2022. ISBN: 978-1-9822-8377-3 (sc), ISBN: 978-1-9822-8378-0 (e)

6. How Did God Create Humankind? Scientific and Biblical Views. Kingdom Publishers, UK 2022. ISBN: 978-1-911697-64-0

Acknowledgement

The photograph of the author was produced by

ZigZag Photography, Leicester UK

studio@zigzagphotography.co.uk

Milton Keynes UK
Ingram Content Group UK Ltd.
UKHW021032260124
436737UK00006B/45